Spiders

101 Fun Facts & Amazing Pictures
(Featuring The World's Top 6 Spiders)

Table of Contents

Introduction

The Mysterious Black Widow

The Dangerous Brazilian Wandering Spider

The Shy Brown Recluse Spider

The Deceptive Crab Spider

The Fun-loving Jumping Spider

The Beefy Tarantula

Conclusion

Introduction

Many people seem to think that spiders are insects. But the truth is, they are not; they belong to the arachnid family. Mites, ticks and scorpions are close relatives of spiders.

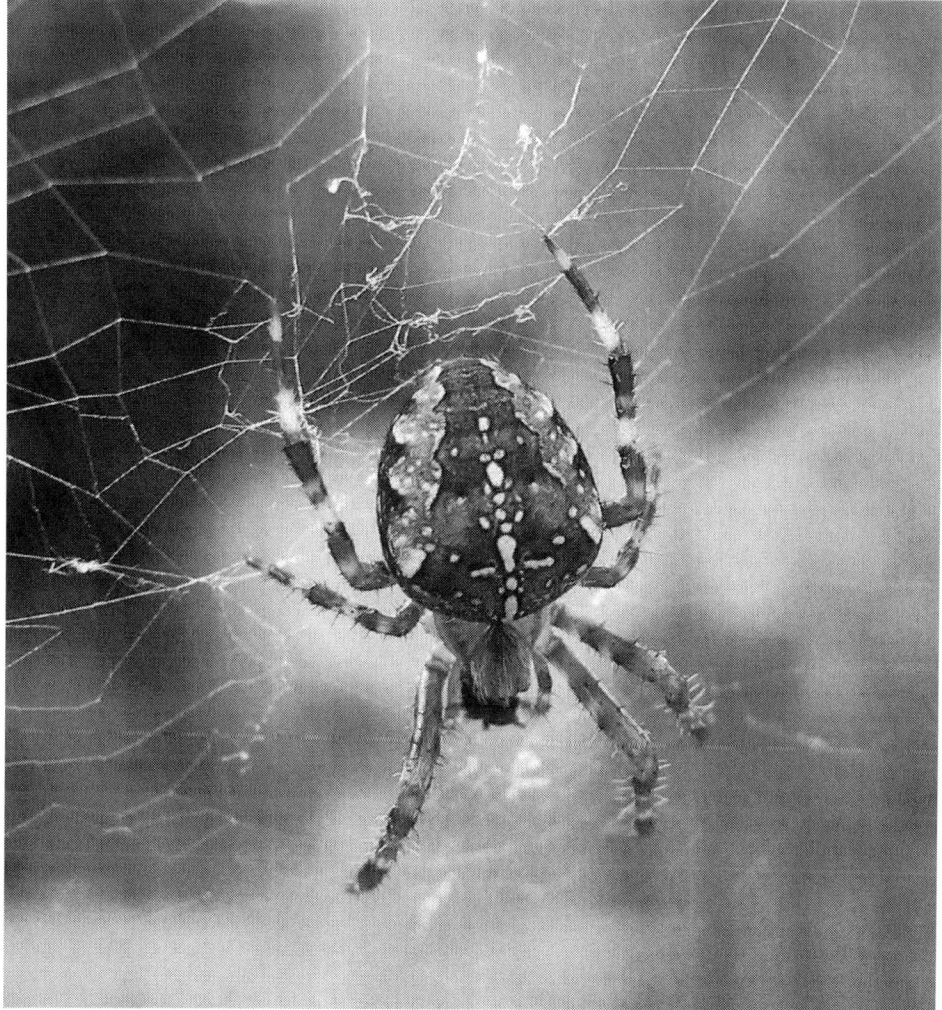

Figure 1: Just another spider hard at work, producing silk and spinning its web.

1. Insects only have 6 legs while spiders have eight.

2. Spiders neither have wings nor antennae but insects do.

3. Spiders cannot chew unlike insects.

4. Like other members of the arachnid family, a spider's body is made of only two segments while insects have three main parts.

5. The two segments of a spider's body include the Cephalothorax and the abdomen. The first segment includes the brain, the eyes, mouth fangs, glands and stomach.

6. Most spiders have eight eyes but other species have fewer eyes.

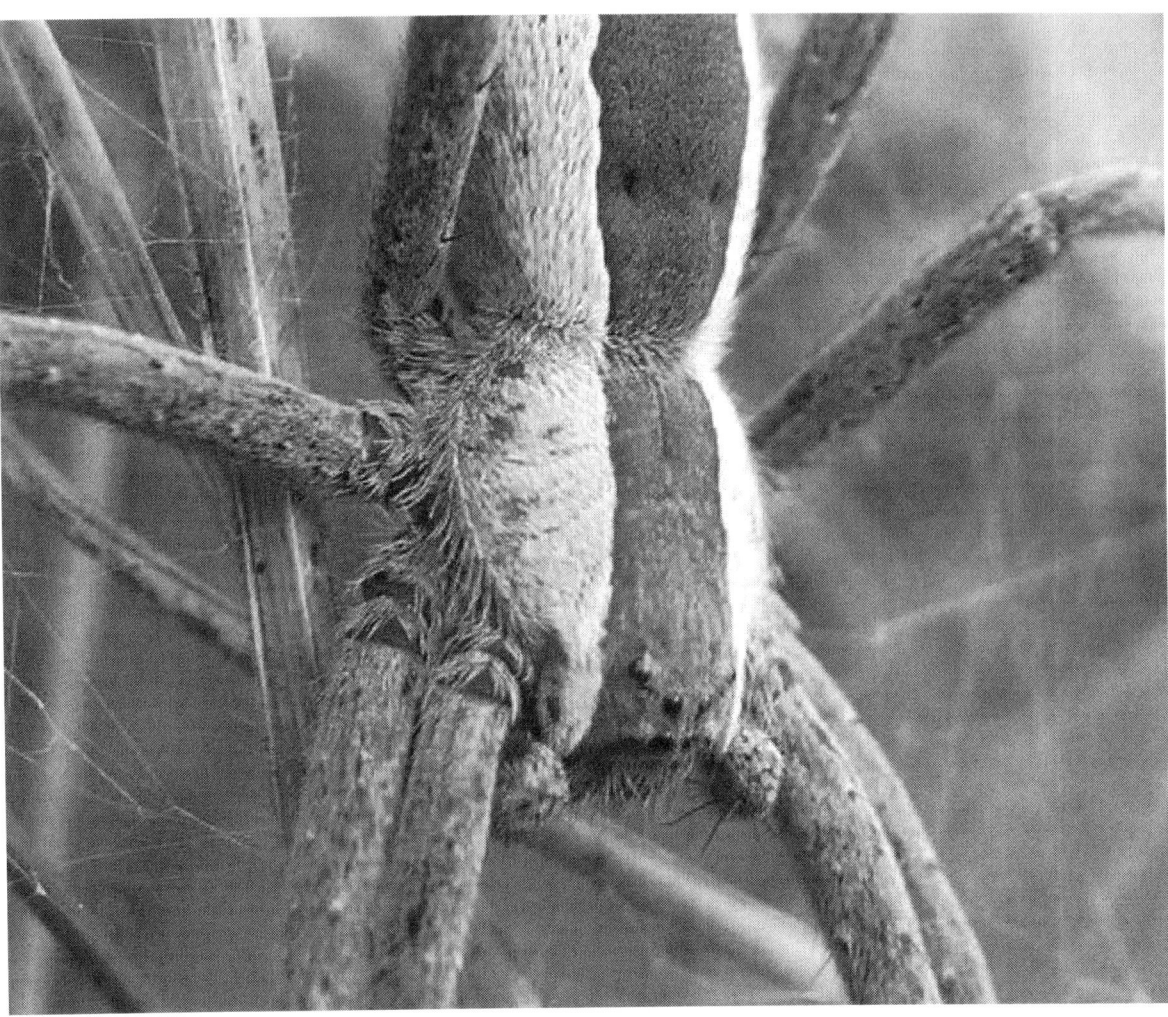

Figure 2: A closer look at a spider

7. Spiders have pedipalps which are leg-like parts located near the fangs. They use this body part to pin their prey down.

8. The second segment of the spider's body, the abdomen, is where the spinnerets are found. These are the glands responsible for producing the silk.

9. Over 30,000 species of spiders exist today.

10. Spiders are mostly predators. They will in fact, eat other spiders. But they usually feed on smaller insects that get caught up in their webs.

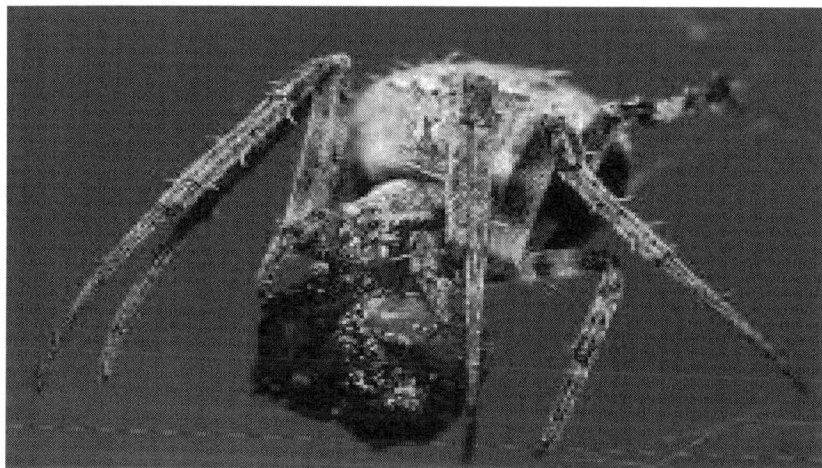

Figure 3: A spider and its meal

11. Scientists have found proof that spiders have been in existence for 2 million years.

12. The spider's skeletons are rather small and quite fragile. This makes it difficult to locate fossilized spiders that are still intact or whole.

13. Instead of a skeleton, a spider has a hard outer shell referred to as the exoskeleton.

14. The exoskeleton is made of a hard material which means the spider is likely to outgrow the exoskeleton.

15. Young spiders shed theirs. Once they have climbed out of the exoskeleton, spiders must stretch out so they have growing room before the new exoskeleton completely hardens.

16. Adults no longer molt because they stop growing.

17. How come the spider does not get stuck on its own web? That's because the spider has some form of oil which prevents it from getting stuck in its own web.

18. Spiders have hairy legs. These small hairs help them out in detecting smells and vibrations from the air.

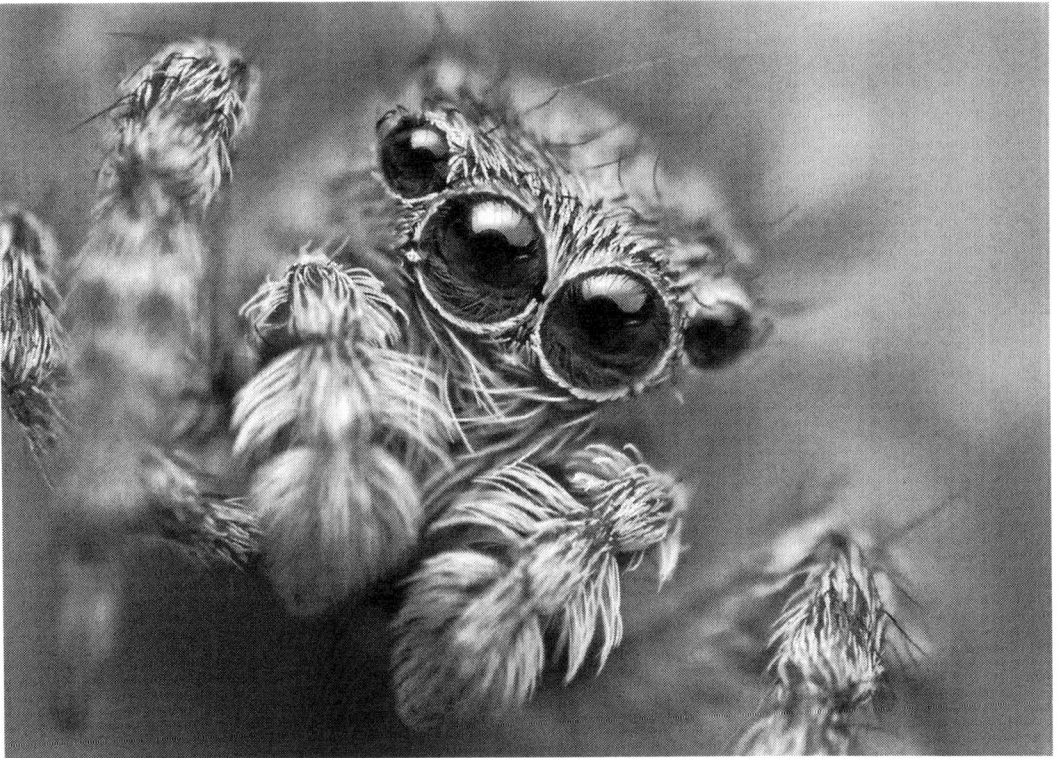

Figure 4: A spider's leg is covered with lots of hair which makes it ticklish to the touch.

19. Spiders have two small claws found at the end of each of its legs.

20. Spiders have a total of 48 knees. They have eight legs with six joints on each.

21. All spiders can produce silk but not all of them can spin webs. Those that don't spin webs are hunters that have their own way of catching food.

22. All spiders have fangs and almost all produce venom. However, there are only a few with venom strong enough to harm humans. Most are adequate in paralyzing their prey but are too mild to affect humans.

23. The smallest spider is the Patu Marplesi. They are so small you can fit 10 on a pencil's end.

24. The biggest spider is the Giant Bird-eating spider whose legs are 11 inches long.

25. Spider silk is considered the strongest material. According to scientists, if spider web is gathered with the same amount and weight as that of a piece of steel, it can turn out to be much stronger.

26. So far, scientists have been unsuccessful in copying spider silk.

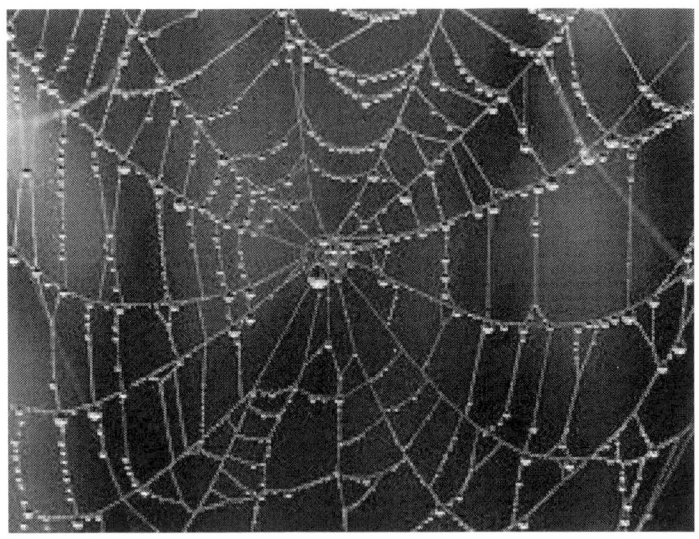

Figure 5: Spider webs are made of fine spider silk.

The Mysterious Black Widow

Found in North America, the Black Widow consists of five species. They are probably the most poisonous spiders found in this region.

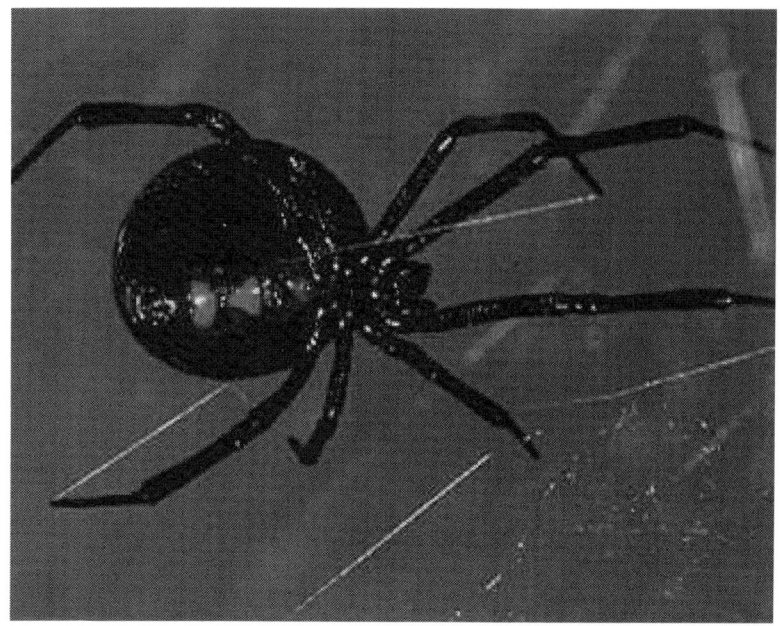

Figure 6: The hourglass red marking on this spider's abdomen spells out poison.

27. Not all Black Widow spiders are venomous. The adult female is what you should avoid.

28. An adult female black widow spider is recognizable with the red hourglass shape found under its abdomen.

29. This poisonous spider is also identifiable because of its shiny black body.

30. The Black Widow spider's body is also made of red markings usually found on its top.

31. This spider can grow to between 8 and 10 millimeters.

32. The Black Widow can lay as many as 400 eggs at a time.

33. The problem is this spider is also a cannibal. That means Black Widow spiders are more likely to kill and eat each other.

34. Female Black Widow spiders have earned a bad reputation as they are thought to always eat their mate. The truth is they do not usually eat their mates. They do so only when they mistake their partners for food.

35. The Black Widow spider is known to produce the strongest silk.

36. These spiders do not bother spinning beautiful webs. Instead, they tend to spin thick jumbles resembling cobwebs.

37. The thick jumble of web is where they catch food including moths, grasshoppers, flies, beetles and other spiders.

38. Among the greatest enemies of the Black widow spider as with other spider species are wasps and the Praying Mantis.

39. Birds eat black widows too but the spider gives them an upset stomach due to the poison.

40. The red markings on this spider's body serve as warning to predators although some will still dare to eat the Black Widow.

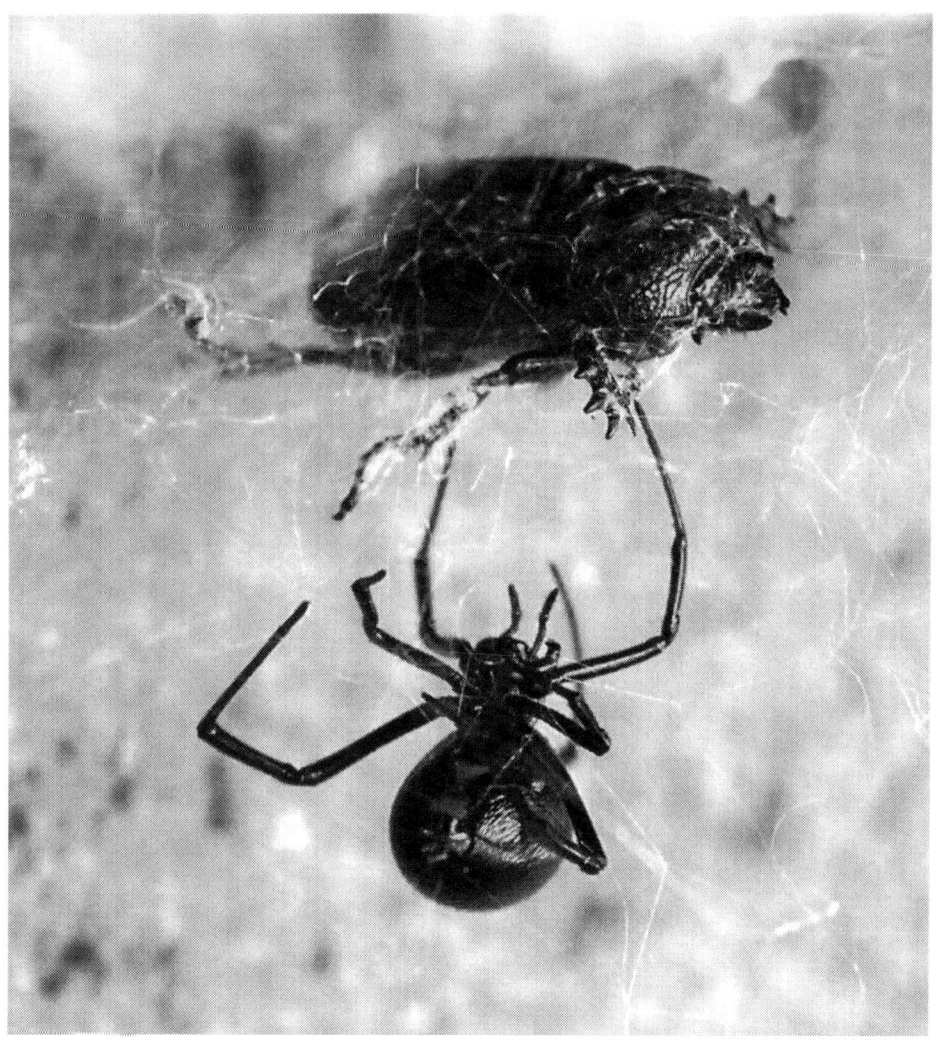

Figure 7: This spider makes a successful catch which means it is meal time!

The Dangerous Brazilian Wandering Spider

This is one of the spiders you should watch out for. The Brazilian Wandering Spider is known as the most dangerous spiders in the world because of its deadly venom.

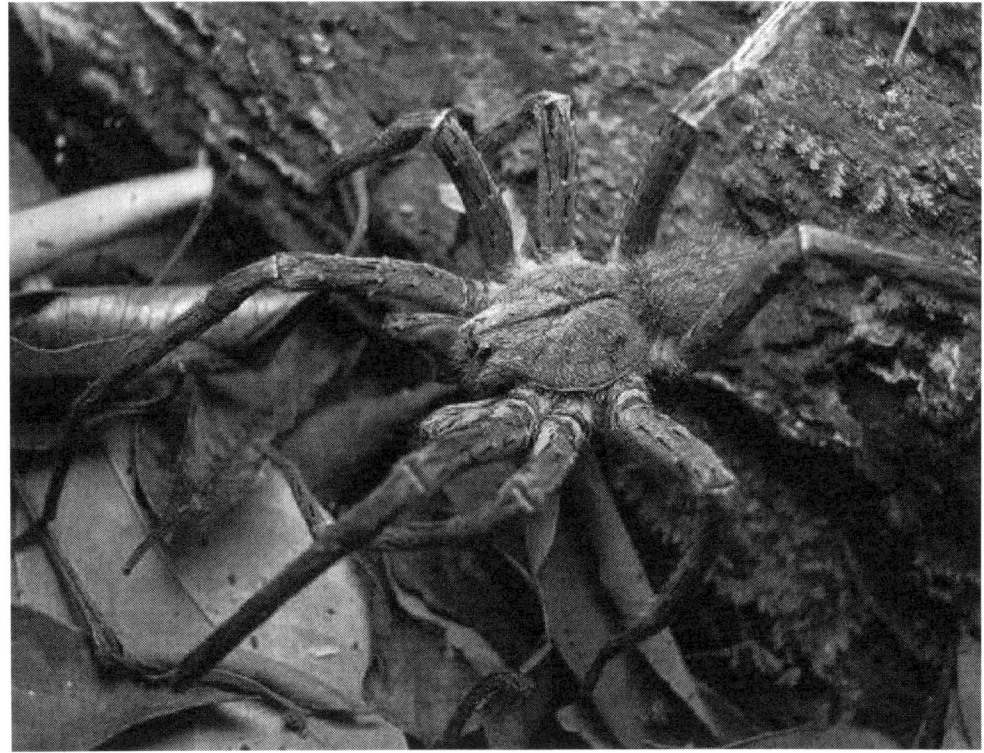

Figure 8: This dangerous spider is quite enormous with long legs that measure up to 6 inches.

41. This spider's legs can grow to as long as 6 inches.

42. Its abdomen has a diameter of around 2 inches.

43. This spider needs to lift two of its front legs up, and would then sway from side to side.

44. The Brazilian Wandering Spider can be found in jungle floors. During daytime, they use crevices and rocks as hiding places.

45. They especially like to hide in dark and moist areas.

46. The Brazilian Wandering Spiders live in the forests of Brazil, Paraguay, Costa Rica, Columbia and Peru.

47. People who live in these regions have to be very careful because these dangerous spiders like to stay in undisturbed items such as piles of wood, garages and closets.

48. They can also be found in banana crates that are shipped all around the world.

49. The Brazilian Wandering Spider is aggressive and will not hesitate to fight for its territory.

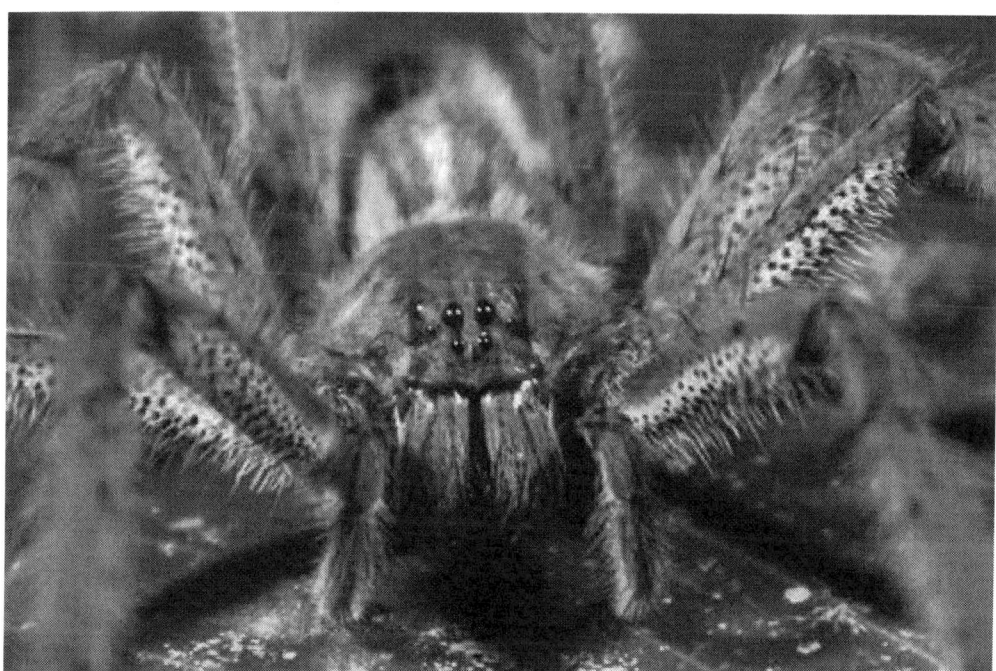

Figure 9: You may not want to get this close to the venomous Brazilian Wandering Spider.

50. They do quite well in the wild. But in captivity, they are most likely to die. They can suffer from a great deal of stress and they have a tendency to stop eating.

51. During the mating season, the male Brazilian Wandering Spider shows its aggressive nature. They are willing to harm each other if that is what it takes to successfully mate with a female.

52. The female selects its mate. But when the mating is over, the male must run away. Otherwise, it is more likely to be eaten by the female.

53. The female Brazilian Wandering Spider creates a sack made from its own silk to protect its eggs.

54. These spiders like smaller prey. That's because it will only take a small amount of venom to take one.

55. They can also feed on a larger sized prey but they often have trouble producing and injecting enough venom to catch larger prey.

The Shy Brown Recluse Spider

As its name suggests, this spider prefer to be left alone. In fact, the Brown Recluse Spider lives alone for most of its life.

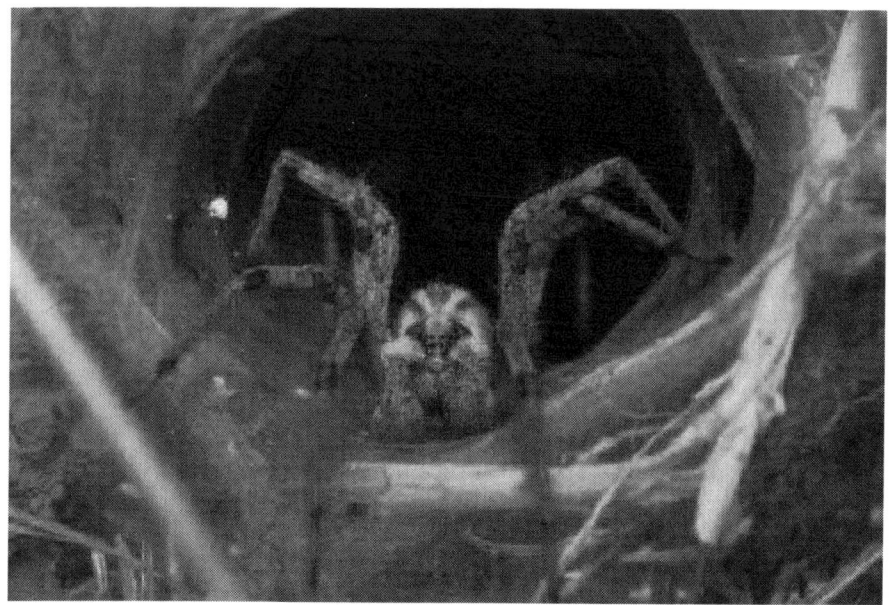

Figure 10: It is wiser to leave this spider alone.

56. The Brown Recluse Spider is usually dark or light brown. Its front part has a violin-shaped dark marking.

57. It has six eyes unlike most spiders that have eight.

58. This spider can be found in the United States, Canada and other parts of the world.

59. You should leave Brown Recluse Spiders alone for a good reason; they are armed with dangerous venom that can harm humans.

60. This spider does its best to avoid human contact. But it will do anything to protect itself too.

61. This spider does not attack without a reason. But when it feels threatened, it is more likely to bite.

62. The Brown Recluse Spider can be found in dark places. They are most likely to hide in closets, garage, shoes, bedding and clothing.

63. This spider prefers to bite and eat insects but it will bite humans when it is disturbed.

64. The Brown Recluse Spider spins sheet webs which it uses to catch insects for food.

65. These spiders are capable of laying eggs for several times in a given year.

66. The Brown Recluse Spider is nocturnal. It usually leaves its hiding place to search for food when it is dark.

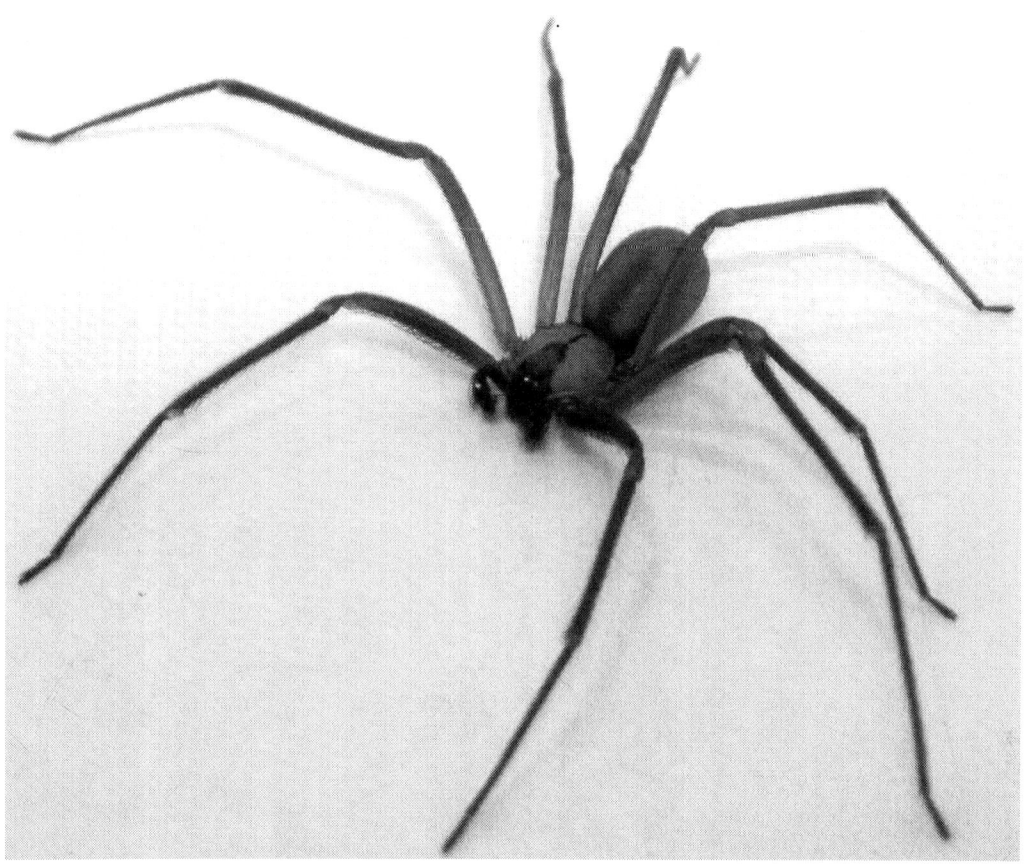

Figure 11: Aside from its long legs and brown color, you will know it is a Brown Recluse Spider when you see that dark, violin-shaped marking on the upper part of its body.

67. A bite from a Brown Recluse Spider will cause the area to swell with white blister. This area is likely to turn red and harden. It is important to seek medical attention immediately if you have been bitten by this spider.

The Deceptive Crab Spider

The Crab Spider can take many forms because of its excellent camouflage skills. This is perfect for disguising itself, patiently waiting for an insect to come near and grab it as fast as this spider can.

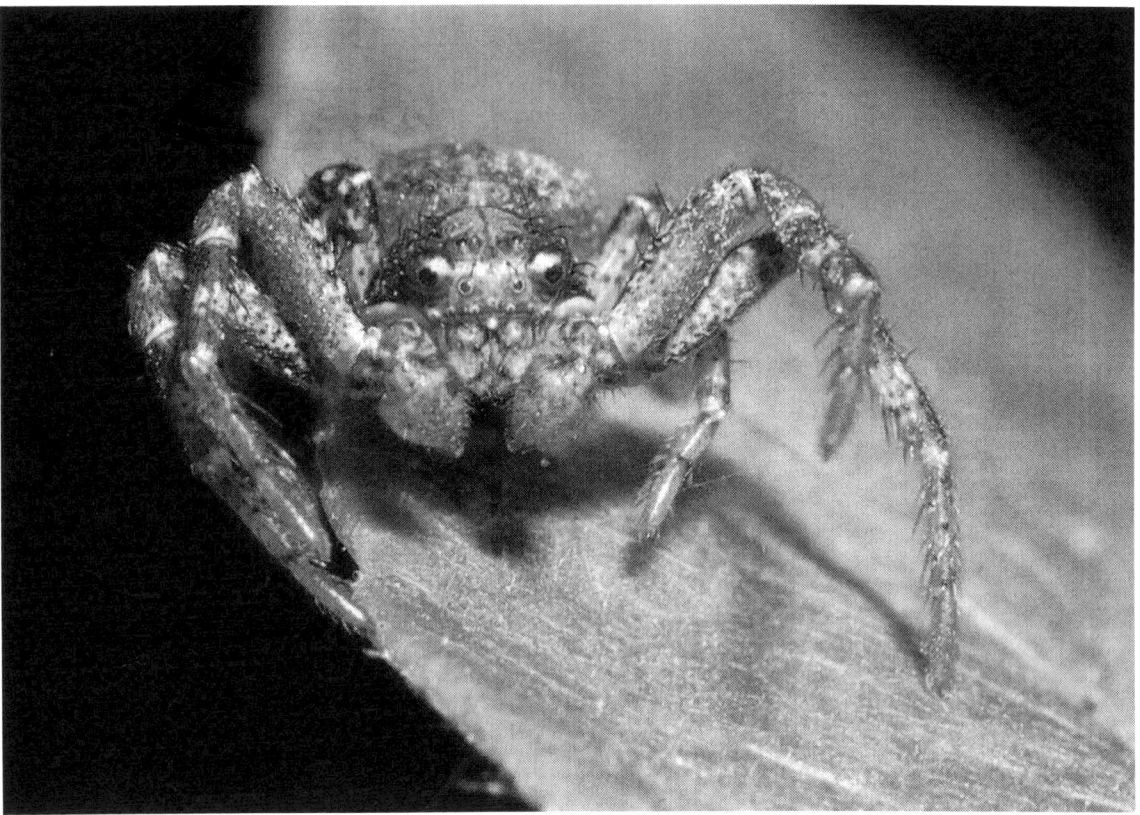

Figure 12: This spider can blend well to its environment literally. It can copy the colors of its hiding places. The camouflage process happens over a period of days.

68. The Crab Spider has a wide, flat and short body.

69. Its first two pairs of legs are much larger than the rest.

70. This spider walks sideways or at times, backwards.

71. Their eight eyes are raised on bumps which allow them to see in every direction.

72. The Crab Spider has small fangs but they are equally venomous. The venom this spider produces can paralyze a prey instantly.

73. Crab Spiders come in a variety of colors. It usually matches their habitat. They can easily match the colors of a flower but it may take them several days.

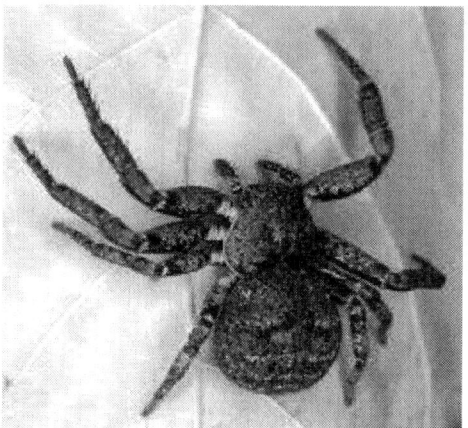

Figure 13: A brown crab spider with black markings

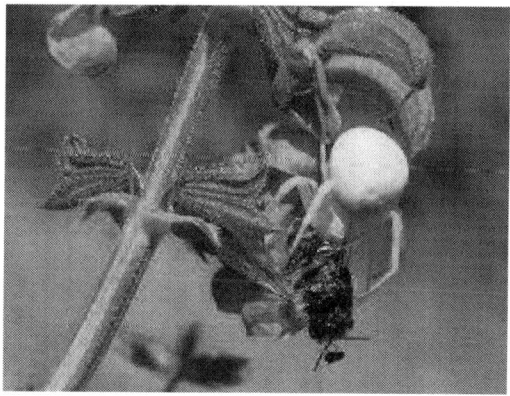

Figure 14: A yellow crab spider

74. Overall, crab spiders have more than 2,000 species. In North America alone, more than 200 species of crab spiders exist.

75. They can adapt to almost any habitat except in dry deserts and cold mountains where there may not be plenty of insects to eat.

76. The female Crab Spider lays the eggs and protects them in a sack made from spider silk.

77. When the spiderlings are hatched, they resemble the shape of adults.

78. As the spiderlings grow, they would shed their skins but their general shape remains.

79. The Crab Spider lives for only one year or less if the climate in the area is cold. However, they can live longer in tropical settings.

80. These spiders are solitary. They live and do the hunting on their own.

81. Crab Spiders hunt for food during daytime.

82. These spiders are not aggressive and they are not considered dangerous to humans.
83. They like to feed on insects and mites.

84. There are many species of crab spiders. The Japanese spider crab is probably the largest with a leg span that measure up to 4 meters.

The Fun-loving Jumping Spider

Jumping spiders come in various sizes and colors. But one thing they all have in common is the way their eyes are arranged. They usually have 3 or 4 rows of eyes. The front row eyes are more forward. The middle two eyes tend to be bigger than the rest.

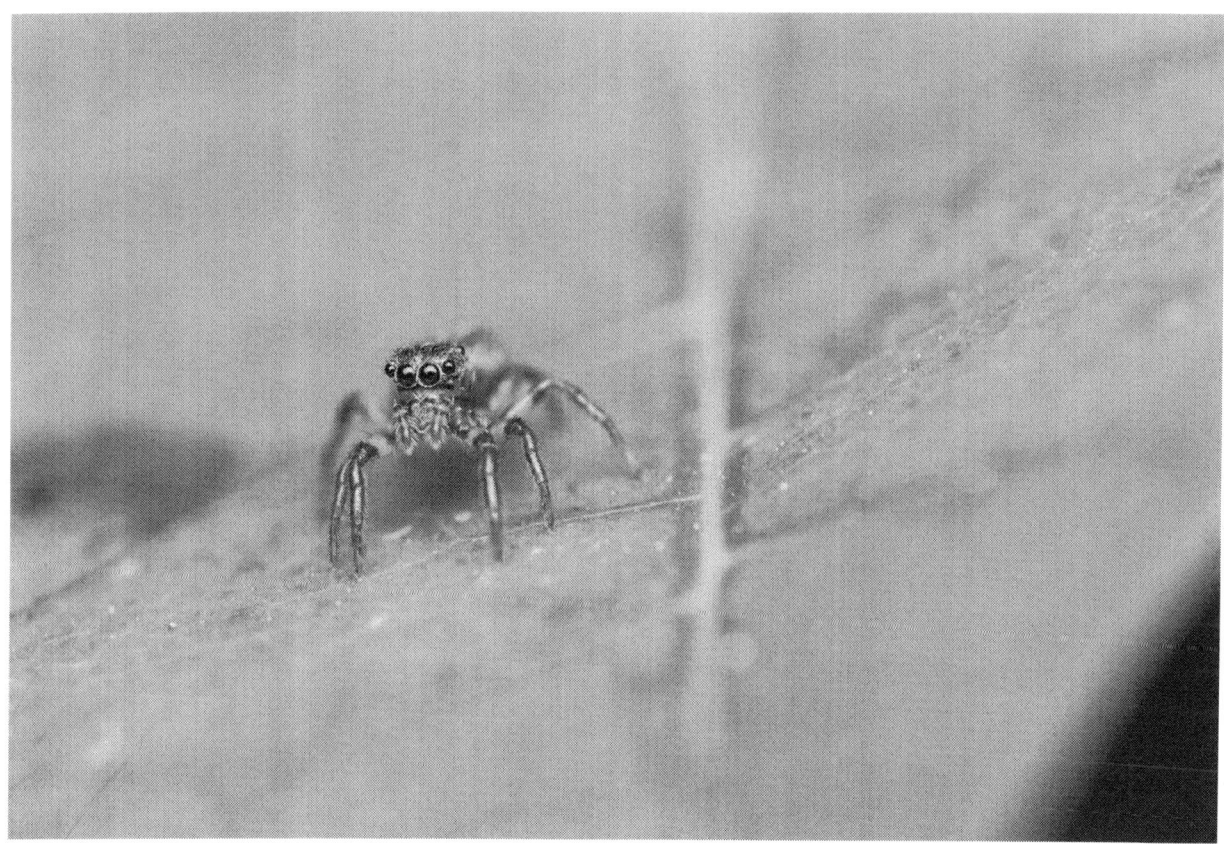

Figure 15: Jumping spiders are recognizable with the way their eyes are arranged, the middle two being larger than the rest.

85. The Jumping Spider family consists of 5,000 species making them the biggest spider family in the world.

86. They can survive in various habitats as long as there are insects to feed on.

87. Male jumping spiders woo the females through courtship dances. The female usually responds by dancing as well and that shall mark the mating.

88. Female jumping spiders can lay hundreds of eggs which they protect in a covering made from their own silk.

89. The mother jumping spider usually guards its eggs but this is not always the case.

90. The newly hatched jumping spiders look similar to the adult spiders.

91. Young jumping spiders molt around 5 or 6 times before they fully grow into adults.

92. Jumping spiders do not spin webs to catch food. Rather, they hunt their food.

The Beefy Tarantula

Some people have tarantulas as pets. But others cringe at the thought of touching one.

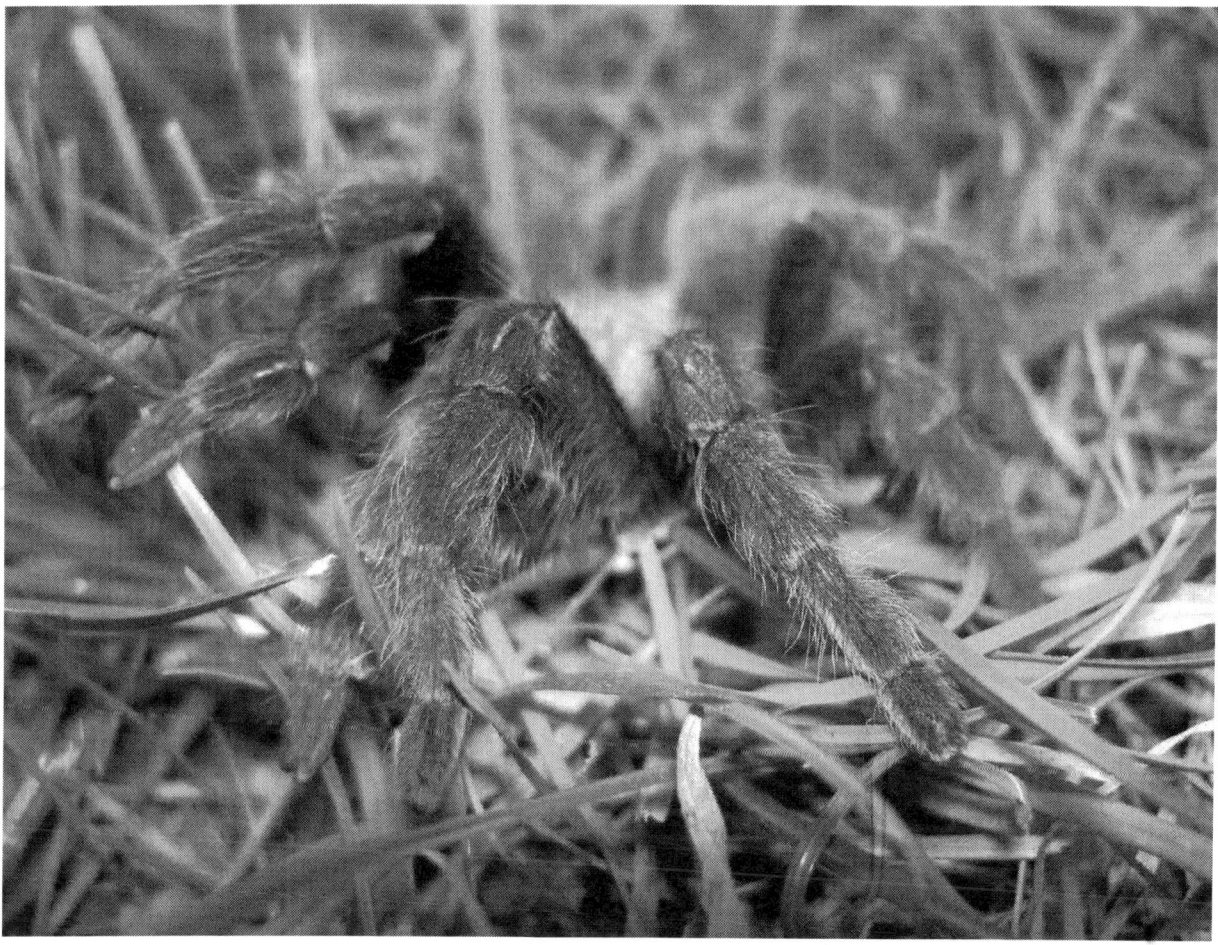

Figure 16: The large and hairy body and legs of a tarantula scare most people away.

93. They can measure up to 4.75 inches with a leg span at 11 inches. They can also weigh as much as 85 grams.

94. Tarantulas may look creepy and their bite may be painful. However, their venom is weak. In fact, a bee sting is way more venomous than that of this spider.

95. They are slow movers that hunt their prey at night time.

96. They like insects but they will also go after toads, frogs and mice.

97. When a large tarantula is treated to a large meal, it can live on for a month without feeding.

98. Tarantulas live in desert regions as well as tropical and subtropical locations.

99. These spiders are burrowers. They do not spin a web but will use their silk to set up a trip wire to alert them in case of an intruder to their burrow.

100. A female tarantula patiently guards its eggs placed in a cocoon. It will take 6 to 9 weeks before they hatch. That can result to 500 or as many as 1,000 tarantulas.

101. These spiders can live as long as 30 years in the wild.

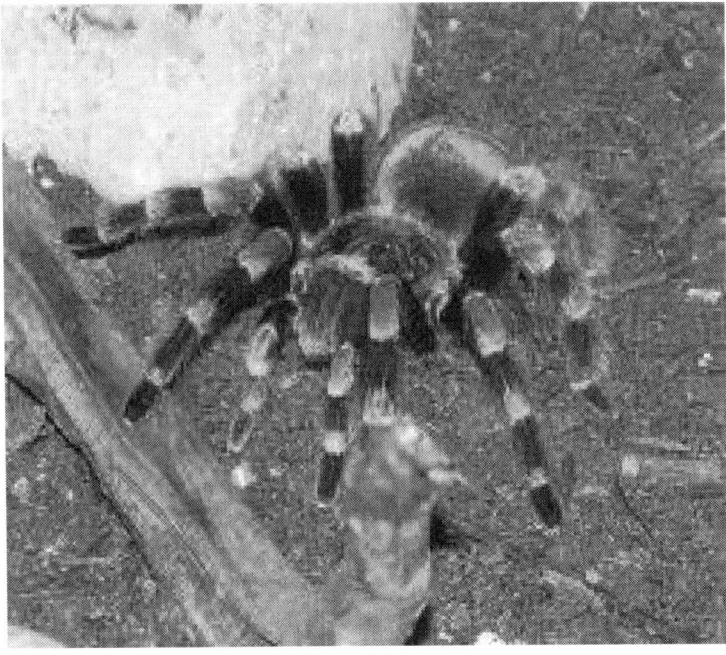

Figure 17: Another tarantula specimen

Conclusion

Spiders are actually helpful because they eat insects and mites. However, most spiders are rather small and that means they have plenty of predators too. That includes birds, lizards, toads and monkeys.

The spider's most notorious enemy is the Spider-wasp. The female wasp can inject venom that causes the spider to become paralyzed. The wasp will then dig a hole where it buries the spider along with its eggs. The spider serves as food to the newly hatched wasps.

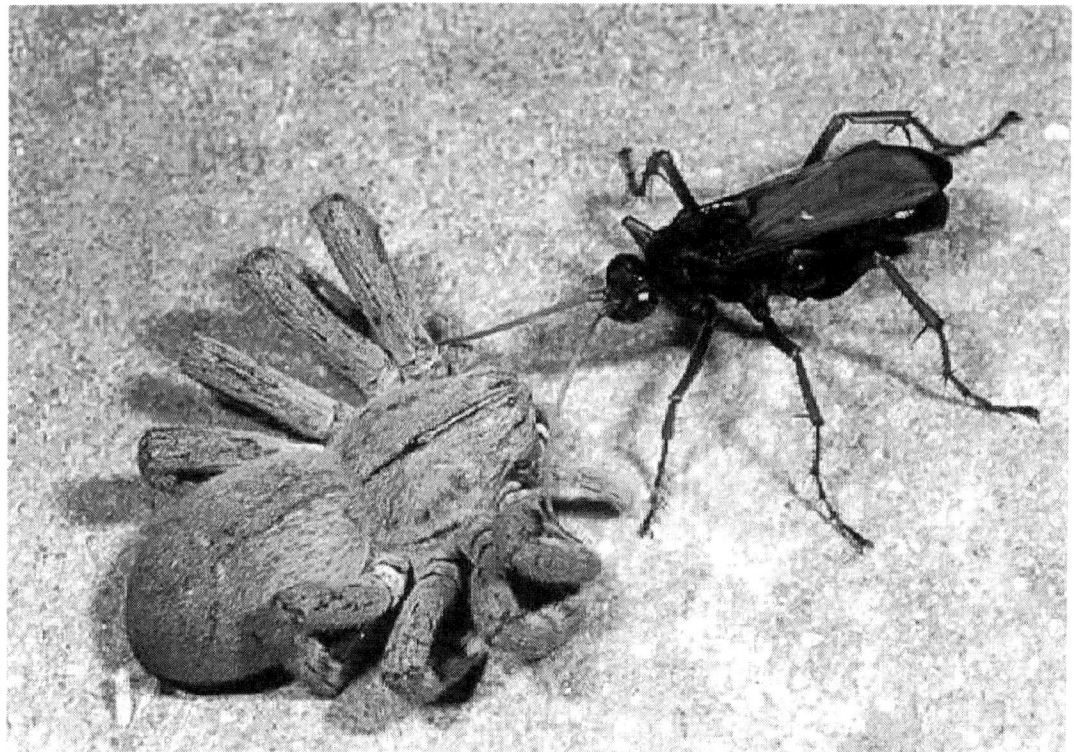

Figure 11: The wasp is one of the spider's most dangerous enemies.

In addition, humans are also a threat. People can kill spiders easily just by stepping on them. Pesticides that humans use also kill many spiders and insects.

Printed in Great Britain
by Amazon